SOMETHING BLUE

SOMETHING BLUE

CHARLOTTE HARRIS

To order additional copies of this book, contact:
Xlibris
UK TFN: 0800 0148620 (Toll Free inside the UK)
UK Local: 02036 956328 (+44 20 3695 6328 from outside the UK)
www.Xlibrispublishing.co.uk
Orders@Xlibrispublishing.co.uk
809133

CONTENTS

CHAPTER ONE

An Introduction

Time Dilations

The time we live in has very few limitations; this is one of them. It begins with the concept of time as two opposing and yet surprisingly similar ideas about how time flows. One idea suggests that time measures stationary aging. An object that does not move is still governed by time and moves through time whilst going nowhere. This concept of time is seen even more accurately if you look at a piece of fruit on your windowsill. After a few days, the fruit will begin to rot, though if the

moisture is drawn away, it is possible to produce dried fruit without very much difficulty.

Then there is the other idea of the flow of time. Here time is associated with motion and moving at speed. We are zipping across the galaxy at a high velocity, and this is the same as the passing of time, forwards and occasionally backwards—which it seldom does, but it is possible, as much of time is reversible, or has symmetry.

These two ideas of time passing cannot exist simultaneously. Either time is measured at the rate of aging and can be measured with an atomic clock, or time is measured at the rate of motion using a light clock. It is always one or the other, but never both together.

Time dilation occurs when travelling at the speed of light. Einstein's theory of relativity states that a subject approaching

the speed of light will become infinitely heavy, so that it takes so much energy to move it that the speed of light can never be reached.

The little known theory of time dilation maintains that at the speed of light each second passes by as if in a scene frozen in time. Time flows at right angles so that it is free to travel at its own speed. Averaging them both out, travel at right angles is

somewhat slower, and the time that passes one freeze at a time should move at the speed of one hour every minute. This is when we hear the chiming of clocks.

CHAPTER TWO

Exponents

Exponents are powers. Another way exponents can be expressed is in numbers raised to a power.

Logarithms can be added as exponents or numbers raised to a power:

$$\log_5 (125) = 3$$

Five to the third power is 125, then x = 3, so that means 3 is the exponent.

The exponent in adding exponents is the exponent raised to a power:

$$5^3 = 125$$

Previously, 3 has been the power or sometimes the exponent.

Here 5^3 is the exponent, the exponent raised to a power, or the exponent raised to a base.

In this case, logarithms are exponents raise to a decimal number power. If logarithms are expressed as exponents, they can be added as the sum of exponents raised to a decimal number power.

Multiplication of numbers raised to a power follows the rule $A^x \times B^x = (AB)^x$ for the same power and $A^x \times A^s = A^{(x+s)}$ for different powers and the same base.

Addition is not multiplication! So the same rules cannot be used. Addition of exponents has its reverse in subtraction. These are the fundamental rules of adding and subtracting exponents.

The addition of exponents has a fundamental rule woven through the equations that follow:

$$a + b = [(b/a) + 1] * a$$

This is the principle that converts an ordinary addition sum into one of multiplication.

Adding Exponents with the Same Base

$$A^x + A^y = [(A^{x-y} + 1)/ A^{x-y}] * A^x$$

$$8^4 + 8^2 = [(8^{4-2} + 1)/ 8^{4-2}] * 8^4$$

$$= 1(1/64)\ 8^4$$

$$4160 = 4160$$

The result is a real number answer.

The solution to $A^x + A^y$ may be given as a formula or real number solution. Then $A^x > A^y$ when $x > y$.

The equation for adding exponents with the same base works for whole numbers of A raised to x and y, and with fractions and also decimal number powers of x and y.

Adding exponents with the same power

$$G^x + A^x = [(G/A)^x + 1] * A^x$$

$$8^7 + 6^7 = [(8/6)^7 + 1] * 6^7$$

$$2377088 = 2377088$$

The result is a real number answer.

The subtraction of exponents has a very similar solution.

$$a - b = [(b/a) - 1] * -a$$

This subtraction provides a real number answer when a > b.

Subtracting exponents with the same base

$$A^x - A^y = [(A^{x-y} - 1)/ A^{x-y}] * A^x$$

$$5^7 - 5^3 = [(5^{7-3} - 1)/ 5^{7-3}] * 5^7$$

$$= [(5^4 - 1) / 5^4] * 5^7$$

$$= (1 - 1/5^4) \, 5^7$$

$$7800 = 7800$$

Subtracting exponents with the same power

$$G^x - A^x = [(G/A)^x - 1] * A^x$$

$$7^5 - 4^5 = [(7/4)^5 - 1] * 4^5$$

$$15783 = 15783$$

Then $G^x > A^x$ when $G > A$.

There are some shortcuts for both addition and subtraction:

$$A^x + A^y = (A^{x-y} + 1) * A^y$$

and

$$A^x - A^y = (A^{x-y} - 1) * A^y$$

But the best equation is yet to come: the adding exponents equation!

The exponents equation

$$A^x + A^y = A^{(x+y)/2} \left(A^{(x-y)/2} + A^{(y-x)/2} \right)$$

This equation is all you will ever need to know about adding exponents!

CHAPTER THREE

Raising Exponents

Raised Exponents are in the form A^x and A^y. We can add A^x and A^y with the help of A^m where m is a small number between x and y. So $x > m > y$. There are two solutions to Raised Exponents, 1st Equation and 2nd Equation. We can choose x and y which are variables and m

is inbetween. This is the First Equation, addition and subtraction:

1st equation

$$A^x + A^y = \left[A^m + \frac{1}{A^{x-m-y}} \right] \times A^{x-m}$$

$$A^x - A^y = \left[(A^m - 1) + \frac{A^{x-m-y} - 1}{A^{x-m-y}} \right] \times A^{x-m}$$

The other equation is slightly different and it is known as the Second Equation. Here again is the addition and subtraction:

2nd equation

$$A^x + A^y = \left[A^{x-m} + \frac{1}{A^{m-y}} \right] \times A^m$$

$$A^x - A^y = \left[(A^{x-m} - 1) + \frac{A^{m-y} - 1}{A^{m-y}} \right] \times A^m$$

Both Equations for $A^x + A^y$ give the same answer in a different form. The the sum can be written in different forms for every m between x and y. The option is then to choose which formula you prefer, or why not write them all out to see how the power is changed and the equation conserved for every +1 increase in m. Experiment with m, and when confident choose the most sensible answer.

Example 1:

1st equation

$A = 8,\ x = 7,\ \underline{m = 3},\ y = 2$

$$A^x + A^y = \left[A^m + \frac{1}{A^{x-m-y}} \right] \times A^{x-m}$$

$$8^7 + 8^2 = \left[8^3 + \frac{1}{8^{1-3-2}} \right] \times 8^{7-3}$$

$$= \left[8^3 + \frac{1}{8^2} \right] \times 8^4$$

$$= 8^3 \left(\frac{1}{64} \right) 8^4$$

$$2097216 = 2097216$$

Example 2:

2nd equation

$A = 8$, $x = 7$, $\underline{m = 6}$, $y = 2$

$$A^x + A^y = \left[A^{x-m} + \frac{1}{A^{m-y}} \right] \times A^m$$

$$8^7 + 8^2 = \left[8^{7-6} + \frac{1}{8^{6-2}} \right] \times 8^6$$

$$= \left[8^1 + \frac{1}{8^4} \right] \times 8^6$$

$$= 8 \left(\frac{1}{8^4} \right) 8^6$$

$$2097216 = 2097216$$

These equations only work with an m within the boundaries of x and y but if x and y are not changed the same answer

m	1st Equation	2nd Equation
3	$8^3(1/8^2)8^4$	$8^4(1/8^1)8^3$
4	$8^4(1/8^1)8^3$	$8^3(1/8^2)8^4$
5	$8^5(1/8^0)8^2$	$8^2(1/8^3)8^5$
6	$8^6(1/8^{-1})8^1$	$8^1(1/8^4)8^6$

can be found with any m!

For A = 8, x = 7, y = 2, m = 3/4/5/6

The best solution, however, can be found with the exception x = m. Try this!

Proof:

$$Z^x + Z^y = Z^{1/2(x+y)}\left[Z^{1/2(y-x)} + \frac{1}{Z^{1/2(y-x)}}\right]$$

work with the right hand side

Let $f(x,y,z) = Z^{1/2(x+y)}\left[Z^{1/2(y-x)} + \frac{1}{Z^{1/2(y-x)}}\right]$

$= Z^{1/2(x+y)}\left[Z^{1/2(y-x)} + Z^{-1/2(y-x)}\right]$

$= Z^{1/2(x+y)}\, Z^{1/2(y-x)} + Z^{1/2(x+y)} Z^{1/2(x-y)}$

Now $A^n A^m = A^{n+m}$

So using this identity

$f(x,y,z) = Z^{x/2 + y/2 + y/2 - x/2} + Z^{x/2 + y/2 + x/2 - y/2}$

$= Z^y + Z^x$

This is a general proof. It is true for all x, y, z that

$$Z^x + Z^y = Z^{1/2(x+y)}\left[Z^{1/2(y-x)} + \frac{1}{Z^{1/2(y-x)}}\right]$$

CHAPTER FOUR

The Wedding

The other day, I was invited to the wedding of my best friend to her best friend and childhood sweetheart. I asked her how I could help with the preparations. "I need something blue," she said.

The traditional rhyme goes "Something old, something new, something borrowed, something blue."

"I don't have anything blue," I mused. "I'll ask around my other friends."

Now, my maths class had organised a bring-and-buy maths fair. Every person was in charge of a stall. I myself had a stall, but when it was quiet I browsed round the other stalls which were run by my mathematician friends.

I went up to Jan on the cake stall. "My friend is getting married. Do you have anything blue?"

"No," said Jan. "I know about Pythagoras, but I don't have anything blue."

I went up to Judy on the flower stall. "My friend is getting married. Do you have anything blue?"

"No," said Judy. "I know about Euler, but I don't have anything blue."

I went up to Louise on the pet stall. "My friend is getting married. Do you have anything blue?"

"No," said Louise. "I know about Dirichlet, but I don't have anything blue."

I went up to Mark on the drinks stall. "My friend is getting married. Do you have anything blue?"

"No," said Mark. "I know about Legendre, but I don't have anything blue."

I went up to Stephen on the marzipan stall. "My friend is getting married. Do you have anything blue?"

"No," said Stephen. "I know about Wiles but I don't have anything blue."

I walked into a charity shop. I said, "I've spoken to Jan about Pythagoras, Judy

about Euler, Louise about Dirichlet, Mark about Legendre, and Stephen about Wiles. Do you have any books on maths?"

"I am sorry," said the shop girl, "but we don't have any books on maths. Would you like some jewellery? You look like you want something blue."

"Yes," I said, "at last! I like Pythagoras, Euler, Dirichlet, Legendre, and Wiles, but I really need something blue."

"This sapphire necklace is a good buy. Is it a gift? I'll gift wrap it for you."

So I bought the sapphire necklace and went home, stopping in at the house of my friend getting married.

"Oh, it's OK now," my friend exclaimed. "My mother gave me this blue bracelet. Isn't it beautiful?"

So I went to my friend's wedding, and she wore the blue bracelet given to her by her mother. I put the sapphire necklace, still wrapped up, in a dusty drawer and forgot about it. Then I searched out one of my maths textbooks and learnt Fermat.

CHAPTER FIVE

Euler

Euler is held to be one of the greatest mathematicians in history and probably the greatest of the 18th century.

Laplace said, "Read Euler, read Euler, he is the master of us all."

Gauss remarked, "The study of Euler's works will remain the best school for the different fields of mathematics, and nothing can replace it."

By 1726 Euler had his first paper in print, an article on isochronous curves in a resisting medium, and another on reciprocal trajectories, and submitted an entry for the Grand Prize of the Paris Academy. The Academy established a system of prizes in 1721 with a major impact on the development of mathematics and science as they directed work towards important areas.

Euler made decisive and formative contributions to the subjects of geometry, calculus and number theory. Euler carried integral calculus to a higher degree of perfection, developed the theory of trigonometric and logarithmic functions, and threw a new light on almost every branch of mathematics.

In 1741, Euler became a member of the Berlin Academy, where for 25 years he

produced many papers, attributing most of them to the St. Petersburg Academy.

In 1748 he developed the concept of a function in mathematical analysis.

Euler's most important contributions were perhaps "Euler's Formula" and "Euler's Theorem". In the first Euler used Euler angles to specify the orientation of a rigid body. In the other we define the exponentials of imaginary numbers in

terms of trigonometric functions. (There is also another version of "Euler's Formula" that gives the values of the Riemann zeta function at positive whole numbers in terms of Bernoulli numbers.)

Here are these two most important equations:

$$\underline{\text{Euler's Identity}}$$

$$e^{i\pi} + 1 = 0$$

$$\underline{\text{Euler's Formula}}$$

$$e^{i\varphi} = \cos\varphi + i\sin\varphi$$

Aren't they elegant? Project Euler is a series of challenging mathematical /computer programming problems that will require a good deal of mathematical insights to solve. Although mathematics will help you arrive at the most efficient methods, practise will be required to solve most problems.

Euler's Number

The constant e, the base of the natural logarithm.

THE END

www.ingramcontent.com/pod-product-compliance
Lightning Source LLC
Chambersburg PA
CBHW021450210526
45463CB00002B/718